FUN WITH GENE

Teacher Edition

N. C. Bailey & N. L. Eskeland

Science2Discover
Del Mar, CA

Book Cover designed by Eric Beck and N. C. Bailey

Illustrations were done with CorelDraw

ISBN 0-9673811-4-2

 First Edition 2001

Science2Discover
P.O. Box 2435
Del Mar, CA 92014-1735

Web site: http://www.science2discover.com

Our sincere thanks to Ms. Debbie Cully, 7th grade science teacher (Solana Beach, CA), for her valuable comments and suggestions.

TABLE OF CONTENTS

INTRODUCTION

Hello! It's Gene again. Are your students ready to have some fun? It is time for them to apply their knowledge about me using the workbook FUN WITH GENE. The workbook complements the textbooks CALL ME GENE and MY NAME IS GENE.

The book is divided into three sections:

1. **Cases to Solve**. Students will be able to solve cases related to genetic diseases, agriculture, and forensic medicine. They will also learn about pedigrees and how scientists use the GeneBank database to identify and/or retrieve DNA sequences. This section is divided into two parts: THE BASICS and BEYOND THE BASICS. The former includes cases that are easy to solve. The latter contains more challenging cases to solve and a family pedigree. BEYOND THE BASICS is for the more daring students.
2. **Hands-on Activities and Games**. In this section, students will learn how to extract DNA from a vegetable and make simple DNA and cell models. A game will show the students how DNA decoding works. A crossword puzzle and a word search are also included.
3. **Exercises**. In this section, students will test their basic knowledge of DNA, RNA, mutations, and cell structure.

I hope that you and your students will have fun. I will certainly have my own fun watching you work on me!

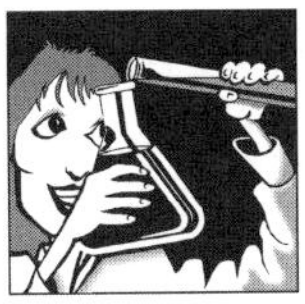

What Makes A Good Scientist?

A good scientist conducts experiments under safe and environmentally appropriate conditions. He/she follows procedures set by his/her institution for conducting research in the laboratory. These include: wearing appropriate clothing, handling all materials cautiously, and disposing of any material according to procedures outlined by the institution. The laboratory should have signs and labels for any hazardous materials and be equipped with instruments that are in good condition. A good scientist conducts experiments with honesty and integrity.

Before a scientist begins with experiments, he/she should first investigate his/her proposal to check its feasibility. He/she should also critically evaluate and analyze the strengths and weaknesses of the proposal by a review of existing scientific literature.

Many grants that scientists apply for from the government require a detailed description of the project. The following are NIH (National Institute of Health) Guidelines for submitting a research proposal:

1. **Specific Aims:** The scientist has to state the specific purpose of his/her proposal and the hypotheses to be tested.
2. **Background and Significance:** The scientist has to write detailed background information on the proposal and state the importance of that

information in relation to his/her specific aims and long-term objectives. He/she should ask questions such as whether the proposal will contribute to society and in what ways.

3. **Research Design and Methods:** The scientist has to write a detailed description of all the procedures that will be used to accomplish his/her aims. These include all instruments and materials that will be used in the experiments.
4. **Preliminary data:** To evaluate the feasibility of a long-term project, many grants require the scientist to perform preliminary experiment(s) before embarking on the project. The scientist must record all his/her observations and findings in a clear and orderly fashion.
5. **Literature Citation:** The scientist has to list all references cited in the research proposal.

SECTION 1

NOTE: In Basics case 1, and Beyond the Basics cases 1-4, the characters depicted are fictitious. Any similarity to existing characters is accidental and unintentional.

THE BASICS

Can you solve the case?

CASE 1

Prerequisite: Punnett Square and basic Mendelian genetics (Chapter 3)

Sickle-Cell Anemia

Bill and Jane were married and a year later, they decided to start a family. They knew that Bill had a great-grandmother who had died from sickle cell anemia. In order to have the disease, a person has to inherit one copy of the gene from each parent. Jane was not sure if any of her ancestors had the disease. When baby Kyle was born, he seemed healthy. Several months later, the baby was not doing well and the doctor diagnosed sickle cell anemia. Bill and Jane were unhappy and confused. They knew that a long rough road was ahead of them. Since both were healthy, they knew that they must be heterozygous for the defective hemoglobin gene. DNA tests on blood samples taken from Bill and Jane confirmed their suspicions.

What do you think are the test results? Let us say that the gene for sickle cell anemia is **s**. Use the Punnett Square and then circle the correct answer from the following:

- Both Bill and Jane have one copy of the **s** gene.
- Jane, but not Bill, has two copies of the **s** gene.
- Bill, but not Jane, has one copy of the **s** gene.
- Both Bill and Jane have two copies of the **s** gene.

Punnett Square:

Answer: Both Bill and Jane have one copy of the **s** gene.

Explain: Each parent must have one copy of the mutated gene in order for the disease to be passed on to a child. The disease for sickle cell anemia **s** is an autosomal recessive trait. Autosomal recessive inheritance means that the gene is located on one of the autosomes (chromosome pairs 1 through 22). Since these parents do not have the disease, they are heterozygous for the disease (**Ss,** where **S** is a normal copy of the gene), the child is homozygous (**ss**).

	S	**s**
S	**SS**	**Ss**
s	**Ss**	**ss**

What is the chance of Bill and Jane's next child developing the disease? Circle the correct answer and explain in the space provided below.

- 100%

- 50%
- 25%
- 0%
- None of the above

Answer: 25%.

Explain: From the Punnett Square, the **ss** genotype is in one quadrant out of the four, which is 25%. Each time Bill and Jane have a child, the probability of the child with the disease will always be 25%.

****

CASE 2

Prerequisite: Mendelian genetics (Chapter 3).

The Pea Plant

In the mid 1800s, Gregor Mendel, a monk known as the "Father of Genetics", planted pea plants in his garden. He first pollinated garden peas that were all tall. The offspring were all tall. He then pollinated garden peas that were short and tall. To his surprise, the offspring were all tall!

From the above experiment, what did Mendel conclude about the phenotype tall? Circle the correct answer and explain in the space provided below.

- It is dominant
- It is recessive
- It is neither dominant nor recessive

Answer: It is dominant

Explain: The tall phenotype is dominant since all the offspring were tall even though the parent plants were tall and short.

Given the above information, what is the genotype of the offspring if one parent is TT (tall) and the other is tt (short)? Use the Punnett Square below to find the answer.

- TT
- tt
- Tt

	T	**T**
t	**Tt**	**Tt**
t	**Tt**	**Tt**

Explain: The Punnett square shows that all offspring have the Tt genotype.

****

CASE 3

Prerequisite: Mitochondrial DNA (Chapter 2)

Mitochondrial DNA (mtDNA) is found in the mitochondria of the cell. Mitochondria are cell organelles in which cell respiration takes place. MtDNA in humans is ~16,000 base pairs in length. The mtDNA encodes proteins necessary for respiration. MtDNA is mainly inherited from the egg cell. The egg cell cytoplasm contains much more mitochondria than the sperm cell cytoplasm.

The Prince, an Imposter or Real?

The following is based on historical facts taken from TIME.com, May 1, 2000, Vol.155, No.18, Science by Nadya Labi.

During the French Revolution, King Louis XVI and Queen Marie Antoinette were imprisoned along with their young son Prince Louis XVII. After his parents' execution, the prince was left in prison to die. Some people thought that the king's supporters rescued the prince and left an imposter in his place.

Centuries later, DNA from the heart of the mummified body of the "imposter" was analyzed and compared to DNA from locks of hair taken from Marie Antoinette and two of her sisters. In addition, DNA was taken from living maternal relatives of the queen for further confirmation.

What organelle of the cell was the DNA extracted from? Circle the correct answer and explain in the space provided below.

- Nucleus
- Mitochondria
- Nucleus and mitochondria
- Ribosome
- All of the above

Answer: Mitochondria.

Explain: Mitochondria contain DNA that is mainly inherited from the mother. The mitochondrial DNA (mtDNA) is very similar from generation to the next.

Forensic scientists did the DNA analysis and the result showed that the "imposter" was indeed Prince Louis XVII, thus ending suspicions about the young prince's short life.

Whose DNA did the "imposter's" DNA match? Circle the correct answer below and explain in the space provided.

- The living maternal relatives, but not Marie Antoinette
- Marie Antoinette and her sisters, but not her living maternal relatives
- King Louis XVI
- The living maternal relatives, Marie Antoinette and her sisters

Answer: The living maternal relatives, Marie Antoinette and her sisters.

Explain: All maternal descendents have almost identical mtDNA.

****

CASE 4

<u>Jesse James: Imposter or Real?</u>

The following is based on historical facts taken from FOX NEWS, May 31, 2000, by Matt Perry, Associated Press.

Jesse James, the famous outlaw who robbed banks and trains during the Civil War years and beyond, died in 1882 by a gunshot wound to the head. Before his death, he had recruited Bob and Charlie Ford to help him rob a bank. However, a $10,000 reward proved too tempting to the Ford brothers. While visiting Jesse at his home, Bob Ford drew his gun and fatally shot Jesse in the back of his head.

There are those who believe that Jesse James faked his death and assumed the name of J. Frank Dalton of Granbury, Texas. They believe that Jesse James lived to be over 104 years old.

In 1995, in Kearny, Missouri, the body of Jesse James was exhumed and DNA was extracted from his teeth. The DNA sample analysis was compared to DNA from the descendants of James' sister. The results showed with 99.7% accuracy that the bones were indeed those of Jesse James.

What type of DNA was analyzed? How did the analysis prove that it was the body of Jesse James in the Missouri grave? Circle the correct answer and explain in the space provided below.

- Genomic DNA
- Nuclear DNA
- Mitochondrial DNA (mtDNA)

Answer: mtDNA.

Explain: MtDNA is highly conserved from one generation to the next through the maternal lineage. Since samples from the sister of Jesse James' (maternal lineage) descendants matched with Jesse's DNA, we can be sure that the DNA samples were mitochondrial. Genomic DNA is nuclear DNA and both parents contribute their genomic DNA equally to their offspring, making it more difficult to trace it to ancestors.

****

CASE 5

X and Y Chromosomes

Beside the 22 pair of chromosomes (called autosomal chromosomes), humans also have two sex chromosomes. Females have the genotype XX, while males have the genotype XY. Mutations in the genes located in the sex chromosomes can cause diseases such as hemophilia.

Hemophilia is a genetically inherited bleeding disorder due to few or no blood clotting factors. People with hemophilia bleed heavily even after superficial wounds. About one third of the cases occur with no previous family history. These cases are probably due to spontaneous mutation in the gene.

Hemophilia is an X-linked genetic condition since it is found in one of the X chromosomes carried by females. Women with a hemophilic gene are called carriers. Each daughter or son of a carrier mother has a 50% chance of being a carrier. Some carriers have low levels of clotting factors, which can cause medical problems, while other carriers do not have any symptoms.

Males who inherit the defective X chromosome have hemophilia since males inherit only one X chromosome. Females must have two copies of the affected allele in order to develop hemophilia. Therefore, the disease is X-linked recessive.

Prince Alexis

The following is based on historical facts taken from "Identification of the Remains of the Romanov Family by DNA analysis" by Peter Gill, Pavel L. Ivanov, Colin Kimpton, Romelle Piercy, Nicola Benson, Gillian Tully, Ian Evett, Erika Hagelberg, and Kevin Sullivan. Nature Genetics 6 (February): 130-135 (1994).

In the early 1900s, Tsar Nicholas and Tsarina Alexandra of Russia had four children: ~~three~~ 4 girls and one boy. Tsarina Alexandra was a carrier of the hemophilia gene, but neither her nor the Tsar had the illness. One of their children had hemophilia. Circle the correct statement and explain in the space provided below.

- One of the girls had hemophilia
- All the children had hemophilia
- Alexis, but not his sisters, had hemophilia

Answer: Alexis, but not his sisters, had hemophilia.

Explain: Alexis inherited the affected X chromosome from his mother. The Y chromosome does not carry the gene for hemophilia.

What is the chance of the Tsar's daughters developing severe hemophilia?

- 100%
- 25%
- 0%
- 50%

Answer: 0.

Explain: Since the X from the Tsar is unaffected; the daughters can never develop the illness.

****

BEYOND THE BASICS

CASE 1

Definition: RFLP (Restriction Fragment Length polymorphism) refers to DNA fragments cut by enzymes specific to a base sequence (restriction enzymes) generating a DNA fragment whose size varies from one individual to another (except for identical twins). RFLP data are used as markers on genome maps and for screening for mutations and genetic diseases. In the case of police investigation, RFLP is referred as DNA Fingerprinting.

A Thief at School!

Early one morning, Mr. Lowry, the janitor, entered the large seventh-grade classroom at Gardenia Junior High to clean the room. With some resistance, he opened the door, which had a wool cap trapped under it. Mr. Lowry kicked the cap behind the door. He scratched his head as he surveyed the room, which looked unusually empty. Moments later, he realized that the four computers were gone; including the hard drives, monitors and keyboards! Shattered window glass caught his eye; he bent closer and saw that several pieces were tainted with blood. He notified the principal, who called 911.

The police arrived within minutes. They barricaded the classroom with yellow police tape and allowed no one into the room until the collection of evidence was complete. One of the police officers picked up the wool cap. Scheduled classes were cancelled for the day and the seventh-graders were jubilant. None of the students recognized the wool cap.

The police decided to send two pieces of evidence for DNA testing. What were they? Circle the correct answer and explain in the space provided below.

- Two computers
- A math book and a history book
- Two broken glass
- A piece of glass and the cap
- None of the above

Answer: A piece of glass and the cap.

Explain: A piece of glass was stained with blood and the cap contained hair. The thief may have gotten hurt while entering the classroom; his blood may be on the glass. In addition, the thief may have left behind his cap. DNA can be extracted from both blood and hair. Books or computers don't usually contain evidence for DNA testing.

The police interviewed shopkeepers across the street who claimed that they had seen three suspicious-looking strangers. They gave the police descriptions of the possible suspects. The police picked up one suspect while he was robbing a house in the neighborhood. Two other suspects were apprehended while they were attempting to steal a car.

At the police station, DNA tests were done on the three suspects, referred to as suspect 1, 2 and 3. One suspects' DNA test result matched the DNA test result obtained from one piece of evidence found in the classroom.

Can you help the police identify the thief? In Figure 1, the RFLP test result on the classroom evidence is labeled EVIDENCE 1. The other RFLP test results on the three suspects are labeled 1 (for suspect 1), 2 (for suspect 2), and 3 (for suspect 3).

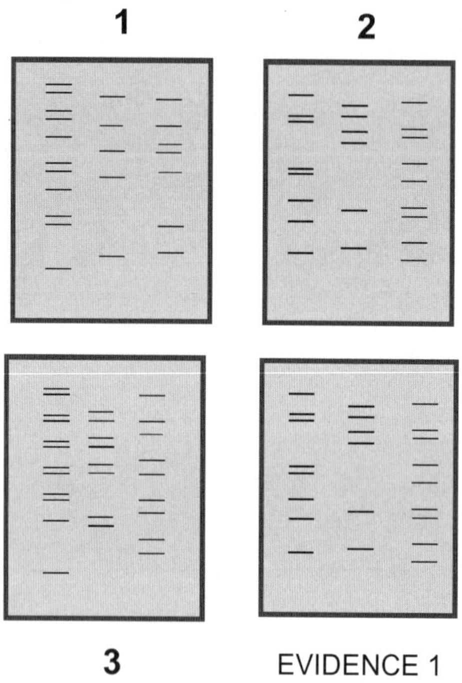

Figure 1

Who is the thief? Circle the correct answer and explain in the space provided below.

- Suspect 1?
- Suspect 2?
- Suspect 3?
- Neither 1, 2, or 3

Answer: Suspect 2.

EXPLAIN: Suspect 2's RFLP profile matches the Evidence 1 RFLP profile.

The police knew that one thief could not have stolen the computers alone. There must have been an accomplice. They hoped that the DNA test done on the second piece of classroom evidence would reveal the identity of the accomplice. The RFLP test result on this evidence is labeled EVIDENCE 2.

Looking at the RFLP from the three suspects in Figure 2, who do you think is the second suspect? Circle the correct answer and explain in the space provided below.

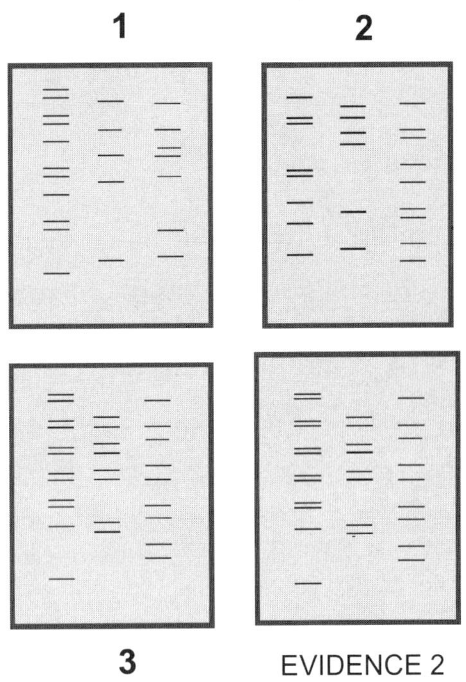

Figure 2

- Suspect 1
- Suspect 2
- Suspect 3

Answer: Suspect 3.

Explain: The RFLP analysis profile for suspect 3 matches that of Evidence 2

CASE 2

Prerequisite: basic cloning technique (Chapter 5)

Definition: Electrophoresis. Cut DNA is run on a special apparatus called a gel electrophoresis that separates the DNA fragments by size. The smaller the fragment is, the further down it will go on the gel.

<u>Where Is the Clone?</u>

Bill, a graduate student, is cloning the gene that controls the formation of wings in birds. He calls the gene "Wing" gene. It is about 500 bp (base pair) long. He inserts the DNA containing the Wing gene into a plasmid (named PLD 3000) that is about 3000 bp long. He goes on a one-week vacation.

Upon his return, Bill goes to the refrigerator to retrieve his newly cloned Wing gene in the plasmid. To his distress, he finds two tubes labeled PLD 3000. One of them must contain his DNA insert. He runs a gel electrophoresis to find out which tube contains his Wing DNA.

Lane 1 shows the molecular size markers (in bp). Lane 2 shows sample A, Lane 3 shows sample B.

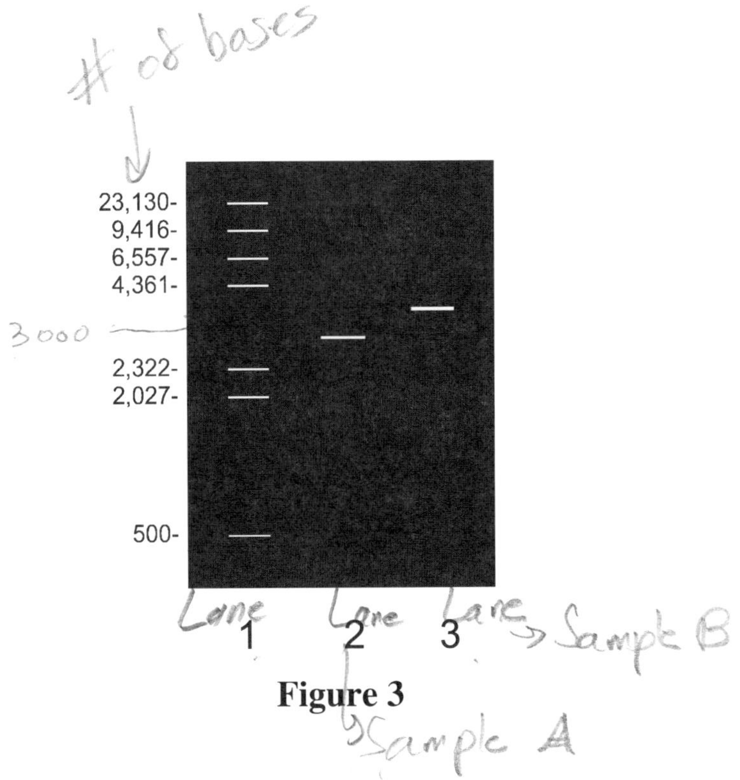

Figure 3

Which sample contains Bill's insert? Circle the correct answer and explain in the space provided below.

- A
- B

How many bases should the new DNA have?

Answer: Sample B.

Explain: It contains the insert since on the gel it is higher than sample A. The size of the plasmid + insert is 3000bp + 500bp = 3500bp. In the gel electrophoresis, sample B ran slower than sample A since it is larger.

****

Pedigree Study

A pedigree is a diagramed list of an individual's ancestors. Pedigrees are used in genetics to determine risk factors for individuals with relatives affected with genetic disorders.

Symbols	Phenotype
square, clear	male, unaffected
square, filled	male, affected
circle, clear	female, unaffected
circle, filled	female, affected
square, half-filled	male, carrier, unaffected or affected
circle, half-filled	female, carrier, unaffected or affected

Vertical lines connect each generation to the next. Horizontal line connects siblings. Roman numerals indicate the different generations while Arabic numerals denote the place of an individual within a given generation. A line linking two symbols shows mating.

CASE 3

Prerequisite: Pedigree for autosomal recessive trait (Chapter 3)

Definition: Carriers. A mutated gene on one of the chromosomes from each parent is required to cause the disease. People with only one mutated gene in the gene pair are called "carriers"; since the gene is recessive, they do not exhibit the disease. Both parents must be carriers in order for a child to have symptoms of the disease; a child who inherits the gene from one parent will be a carrier. Examples of genetic diseases are sickle cell anemia and cystic fibrosis. (Cystic fibrosis clogs the lungs and other organs with sticky mucus that interferes with breathing and digestion. Scientists have found a specific gene mutation involved in most cases of cystic fibrosis).

An Unexpected Gene

Matt and Mary have four children and eight grandchildren. Matt is a carrier for cystic fibrosis disease. One of their daughters is also a carrier for cystic fibrosis. She has five children; one of them, Lydia, has cystic fibrosis. This puzzles the family because Lydia's dad (individual II-6) is not aware of the illness in his family. Everyone is normal. Nevertheless, the family genetic counselor suggests that Lydia's dad undergo a DNA analysis for cystic fibrosis.

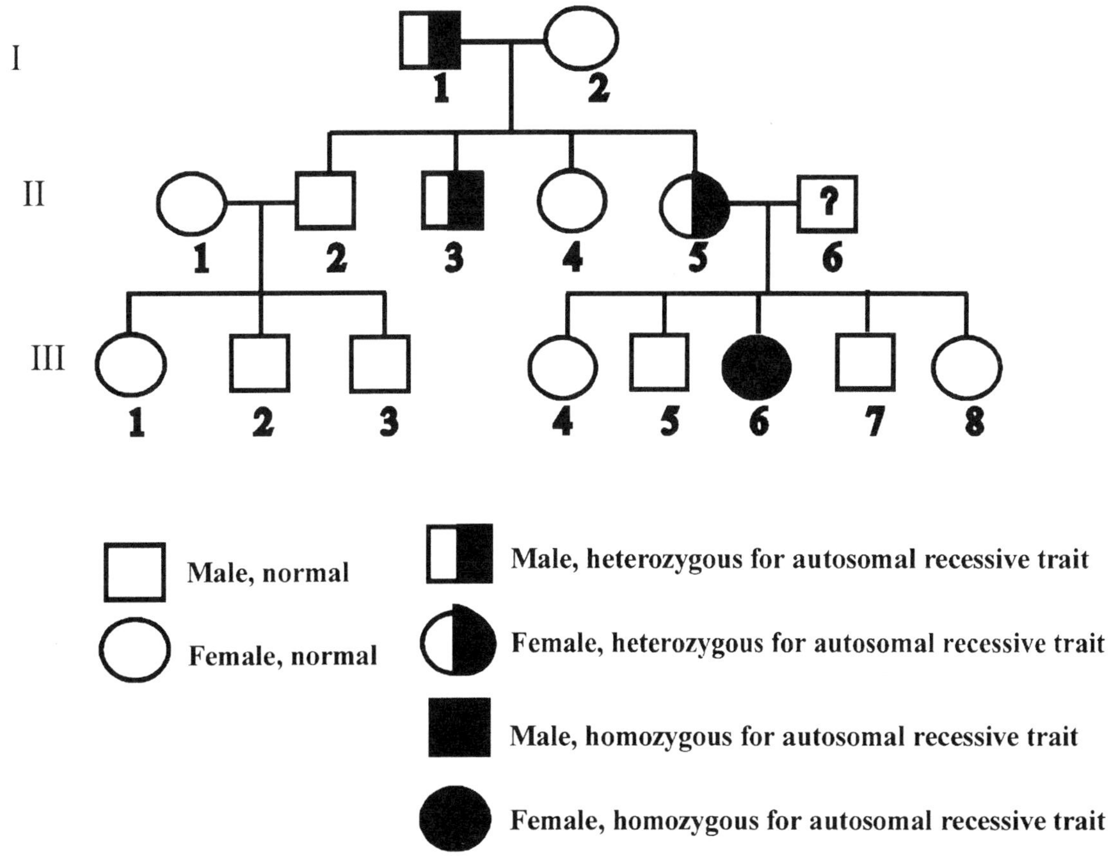

Figure 4

What do you think the result of the test reveals? Circle the correct answer and explain in the space provided below.

- II-6 is a carrier for cystic fibrosis
- II-6 has the disease
- II-6's genotype is normal

Answer: II-6 is a carrier for cystic fibrosis.

Explain: III-6 has the disease. This means both parents, although they are unaffected, are carriers of the cystic fibrosis gene.

Based on your above answer, how do you think the square of individual II-6 should look? Circle the correct answer and explain in the space provided below.

- Filled square
- Half-filled square
- Clear square

Answer: Half-filled square

Explain: Since individual II-6's child has the disease, he must be a carrier.

****

CASE 4

Prerequisite: (Chapter 1, 2 and 3)

A Boy or a Girl?

Ben and Matilda have three sons. They wish to have a daughter. Matilda becomes pregnant. What is the probability that her next baby will be a girl? Circle the correct answer and explain in the space provided below.

- 25%
- 50%
- 60%
- 100%

Answer: 50%.

Explain: Since there are only two choices, a boy or a girl, the probability is always 50% for either sex. It is independent of the order of birth.

Ben and Matilda want to know if they are going to have a girl or a boy. Matilda undergoes a procedure called amniocentesis that allows parents to find out the sex of the child with 100% accuracy. Ten days later, Matilda and Ben are thrilled when they are told that they are going to have a girl.

What did the amniocentesis technician see under the microscope to accurately determine if Matilda is carrying a boy or a girl? Circle the answer and explain in the space provided below.

- DNA
- Gene
- Chromosomes
- Bow ties
- DNA and gene

Answer: Chromosomes.

Explain: Karyotyping (the determination of chromosome number)

What was the sex gene combination for the fetus? Circle the correct answer and explain in the space provided below.

- XY
- XX

Answer: XX

Explain: XX is the genotype for the female.

****

CASE 5

In Search of the Gene

You, a research scientist, just sequenced a gene whose RNA was extracted from the diseased pancreas of a man who has been taking a new test drug for diabetes. The RNA was found to be abundant in this particular organ whereas, previously, the RNA levels for that gene in the man's pancreas were zero. You are cautiously optimistic about the new drug.

The sequence is (the numbers are the positions of the nucleotides):

1 GCTGCATCAG AAGAGGCCAT CAAGCACATC ACTGTCCTTC TGCCATGGCC CTGTGGATGC **61** GCCTCCTGCC CCTGCTGGCG CTGCTGGCCC TCTGGGGACC TGACCCAGCC GCAGCCTTTG 121 TGAACCAACA CCTGTGCGGC TCACACCTGG TGGAAGCTCT

CTACCTAGTG TGCGGGGAAC **181** GAGGCTTCTT CTACACACCC AAGACCCGCC GGGAGGCAGA GGACCTGCAG GTGGGGCAGG **241** TGGAGCTGGG CGGGGGCCCT GGTGCAGGCA GCCTGCAGCC CTTGGCCCTG GAGGGGTCCC **301** TGCAGAAGCG TGGCATTGTG GAACAATGCT GTACCAGCAT CTGCTCCCTC TACCAGCTGG **361** AGAACTACTG CAACTAGACG CAGCCCGCAG GCAGCCCCCC ACCCGCCGCC TCCTGCACCG **421** AGAGAGATGG AATAAAGCCC TTGAACCAGC

To further confirm the identity of the gene that you have just sequenced, you would like to compare your sequence to other known genes. Here is a way to make your comparison.

- Go to the website http://www.ncbi.nlm.nih.gov/ (National Center for Biotechnology Information)
- Click on **BLAST** in the menu. BLAST (Basic Local Alignment Search Tool) is a set of similarity search programs designed to explore all of the available sequence databases.
- Under the heading **Nucleotide BLAST**, click on Standard nucleotide-nucleotide blast [blastn].
- Enter the sequence (without the numbers) above in the **search** box.
- In the **choose database** section, use default **nr**.
- Click on **BLAST!** to start search.

It will take several seconds for the site to complete the search.

Answer: The sequence above is the human insulin mRNA.

SECTION 2

HANDS-ON ACTIVITIES AND GAMES

Crossword Puzzle

Across

2. Immunization
4. A molecule that is the building structure of protein
9. A base
11. Known as "Father of Genetics"
12. Two identical alleles
13. Genetic composition
15. A type of Bacteria
20. A constricted region of a chromosome
22. The science of criminal investigation
27. A signaling protein
29. Discovered DNA structure
32. Ribonucleic Acid
34. Capital of France
36. Island
37. A pouch
39. Not out
40. Codes for amino acid
41. A cord
42. Gaze
43. DNA to RNA
44. To store
45. Gravel
46. Growl
47. A particle
49. A Roman emperor
51. The study of the gene
52. An allele that is not expressed
54. Where a scientist works
56. Symbol, secret language
57. Nourishment
59. Remain
60. Coupling
61. Result of mixing genes or crossing-over
62. Scientific manipulation of living organisms

Down

1. An alternate form of a gene.
3. Rod-shaped body that carry the genes
5. Life form
6. An allele that is expressed as opposed to silent
7. A method of treatment
8. Grease
9. Opposite strand
10. The first cloned animal
14. A reproductive cell
16. Restriction Fragment Length Polymorphism.
17. DNA
18. Without fertilization
19. Indicator
21. An instrument used to see cells and small particles
23. Identical
24. A type of cell division
25. Another type of cell division
26. RNA to protein
28. A structure that protects the cell
30. Protein used to cut DNA
31. Base found only in RNA
33. A type of bond
34. A trait
35. RNA
38. A wrong nucleotide or amino acid
39. To pass a gene from one generation to the next
48. A type of mutation
50. Normal gene as opposed to mutant
53. A microorganism
55. DNA doubling
58. Discovered DNA structure

Answer Key to Crossword Puzzle

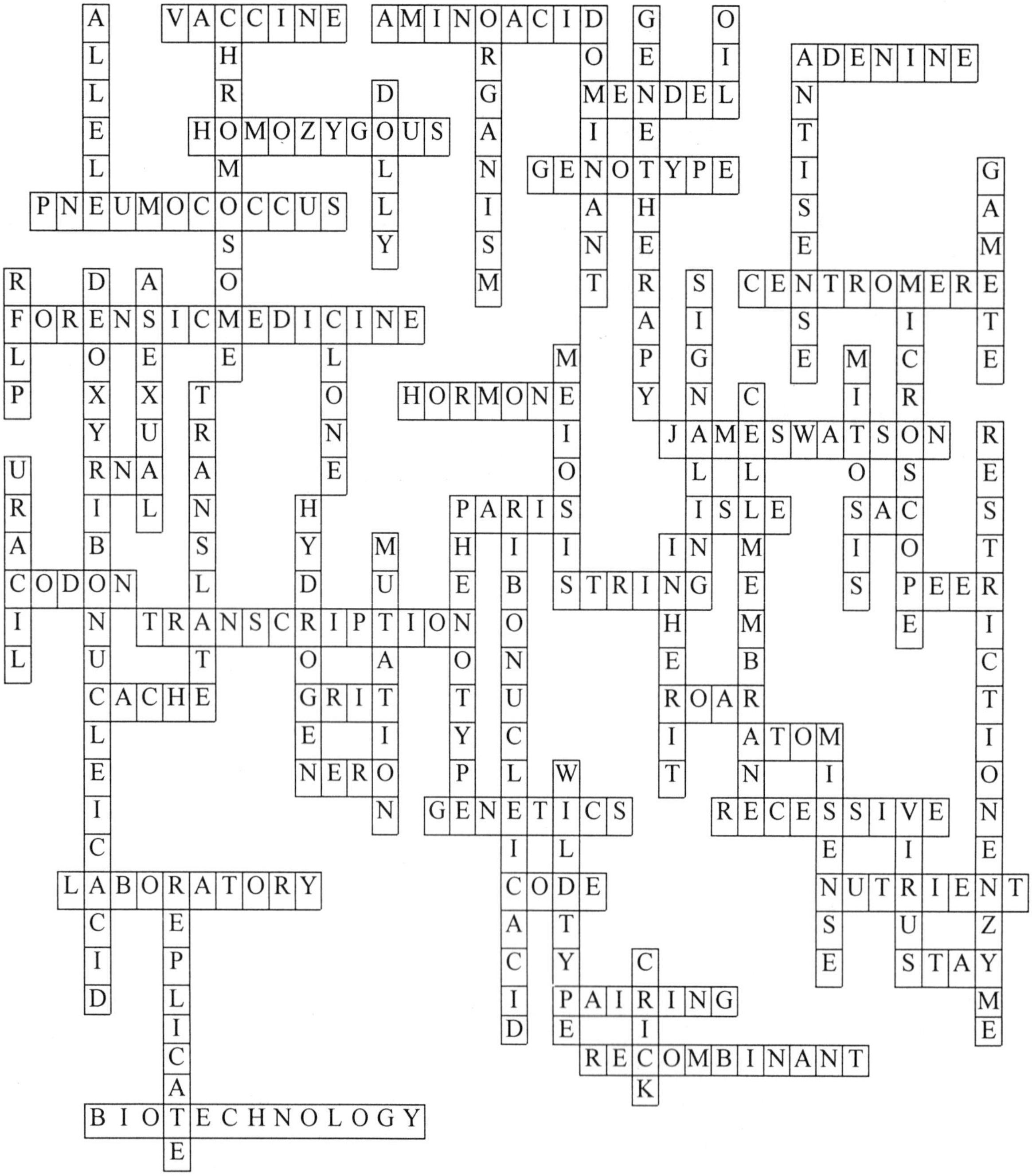

Word Search

A D V A N T A G E I G E N E T I C S G

I C N W R L V L D G T O B F S S N N G

L L E C E L K C I S I B R E E D I N G

I C M B P P E C C T I A D X I N S Y P

H E A I R E T C A B N I U N O N E T R

P R L L O R C N C C T A S L I A D S O

O O A G D C I H I P L E C W S E I A J

M M R C U C S S E S R X T T O S T E E

E N I T C S C P L T O P R O O E O M C

H S A A T R Y R C I E E I R R I E C T

P C V D I L E G U C E R B O E D L L A

N R E C O M B I N A T I O N L O C G R

C E K P N U F A O T F M S Z C B U E E

C E G O I O T C B C G E O I Y I N C G

N N G O R I R O I T N N M O C G R L N

G I A E R U C T R I A T E E L L O C E

N N I E N I S O T Y C O S I L O I U S

R G H O E Y P A R E H T E N E G A E S

N N T E C H N O L O G Y C O C S E E E

I C R O S S I N G U N E N I C I D E M

GENETICS
BREEDING
ADVANTAGE
GOLGI BODIES
RECOMBINATION
RIBOSOMES
FRANCIS CRICK
NUCLEOTIDES
CYTOSINE
BACTERIA
CELL CYCLE
POLYPEPTIDES
MESSENGER
REPRODUCTION
SEXUAL
HETEROZYGOUS
CROSSING
EXPERIMENT
INHERITANCE
TWINS
CYSTIC FIBROSIS
SICKLE CELL
CLONING
SCREENING

HEMOPHILIA
MALARIA
GENE THERAPY
TECHNOLOGY
VACCINATION
RIBONUCLEIC ACID
INSERT
YEAST
PROJECT
MEDICINE
CURE

Answer Key to Word Search

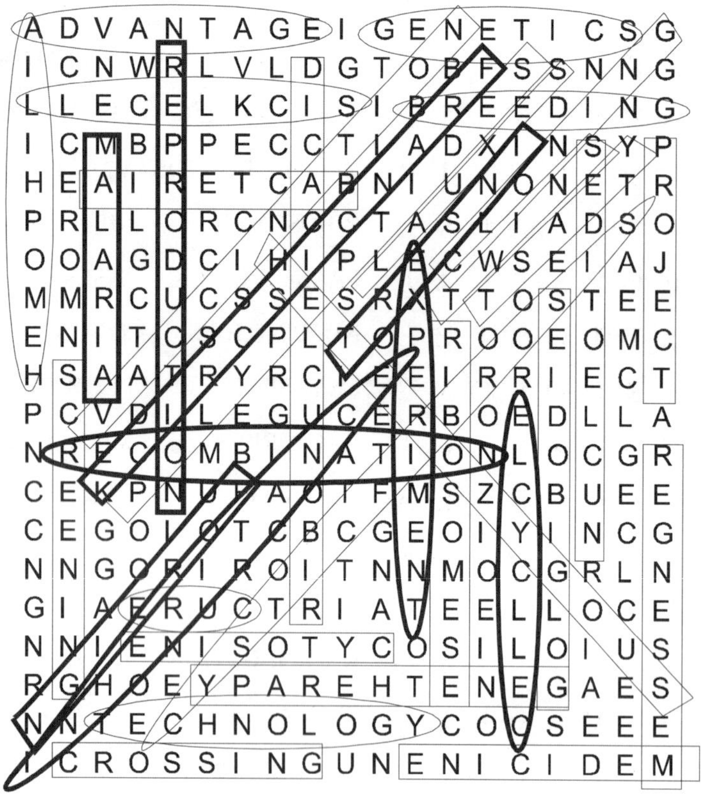

GENETICS
BREEDING
ADVANTAGE
GOLGI BODIES
RECOMBINATION
RIBOSOMES
FRANCIS CRICK
NUCLEOTIDES
CYTOSINE
BACTERIA
CELL CYCLE
POLYPEPTIDES
MESSENGER
REPRODUCTION
SEXUAL
HETEROZYGOUS
CROSSING
EXPERIMENT
INHERITANCE
TWINS
CYSTIC FIBROSIS
SICKLE CELL
CLONING
FOREIGN
SCREENING
HEMOPHILIA
MALARIA
GENE THERAPY
TECHNOLOGY
VACCINATION
RIBONUCLEIC ACID
INSERT
YEAST
PROJECT
MEDICINE
CURE

Decoding the Message

This game is adapted from the video "Genetics" (Discovery Channel). Amino acids make up the backbone of all proteins. Each amino acid is encoded by three nucleotides. There are 20 different amino acids.

Nucleotides (composed of a base molecule, a sugar molecule and a phosphate group) are abbreviated as follows: Adenosine (A), Thymidine (T), Cytidine (C) and Guanosine (G).

For example:
Amino acid Alanine is encoded by any of these triplets: GCT, GCC, GCA, GCG
Amino acid Cysteine is encoded by either of these two codons: TGT, TGC
Amino acid Glycine is encoded by either of these two codons: GGT, GGC
Amino acid Valine is encoded by any of these triplets: GTT, GTC, GTA, GTG
Amino acid Tyrosine is encoded by either of these two codons: TAT, TAC
Amino acid Isoleucine is encoded by any of these triplets: ATT, ATC, ATA
Amino acid Asparagine is encoded by either of these two codons: AAT, AAC
Amino acid Serine is encoded by either of these two codons: AGT, AGC

A coding DNA region will be something like this:
5'-GCTGTTTATGCAGGCGTAGGC-3'
DNA is transcribed to RNA:

5'-GCCUACGCCUGCAUAAACAGC-3'

Translates to a **protein X**:

Alanine-tyrosine-alanine-cysteine-isoleucine-asparagine-serine

THE GAME

Aim: Discover the message encrypted in the colors.

Materials: 4 colored pencils in yellow, red, blue and green; paper.

Procedure: Each letter in the alphabet is encoded by three different colors. Each color is a nucleotide. There are four colors used in this game. Three colors make a codon. The letters are the amino acids. Once you put the letters together, you will get a word and eventually the encrypted sentence! The words and sentences are the proteins.

Put the letters together to find the sentence. Do this in pairs.

Letter codes

A - yellow, green, blue

B - yellow, green, green

C - yellow, yellow, red

D - yellow, yellow, blue

E - yellow, yellow, yellow

F - yellow, blue, blue

G - yellow, blue, red

H - yellow, red, red

I - yellow, green, yellow

J - yellow, green, red

K - yellow, red, green

L - yellow, red, blue

M - red, red, red

N - red, red, blue

O - red, blue, blue

P - red, blue, green

Q - red, green, green

R - red, green, yellow

S - red, yellow, yellow

T - red, blue, yellow

U - red, red, yellow

V - red, yellow, green

W - red, yellow, blue

X - blue, blue, blue

Y - blue, blue, green

Z - blue, green, green

Puzzle Number 1

First Step: Draw two vertical straight lines, about one inch apart, along the length of paper. (You will be provided with a page already set up for one of the puzzles. You will need to provide your own paper for the rest). The two vertical lines represent the primary and complementary DNA strands.

Second Step: Connect both strands with colored horizontal lines. The horizontal lines represent the base pairs connecting both strands. Follow the order of the color guide below.

Yellow→yellow→blue→yellow→green→blue→yellow→yellow→blue→yellow→red →red→yellow→green→blue→red→yellow→yellow→yellow→green→blue→yellow→ green→green→yellow→green→yellow→yellow→blue→red→red→red→blue→red→ blue→blue→red→yellow→yellow→yellow→yellow→yellow.

Connect both vertical lines with colored horizontal lines according to the order that is given to you for each puzzle.

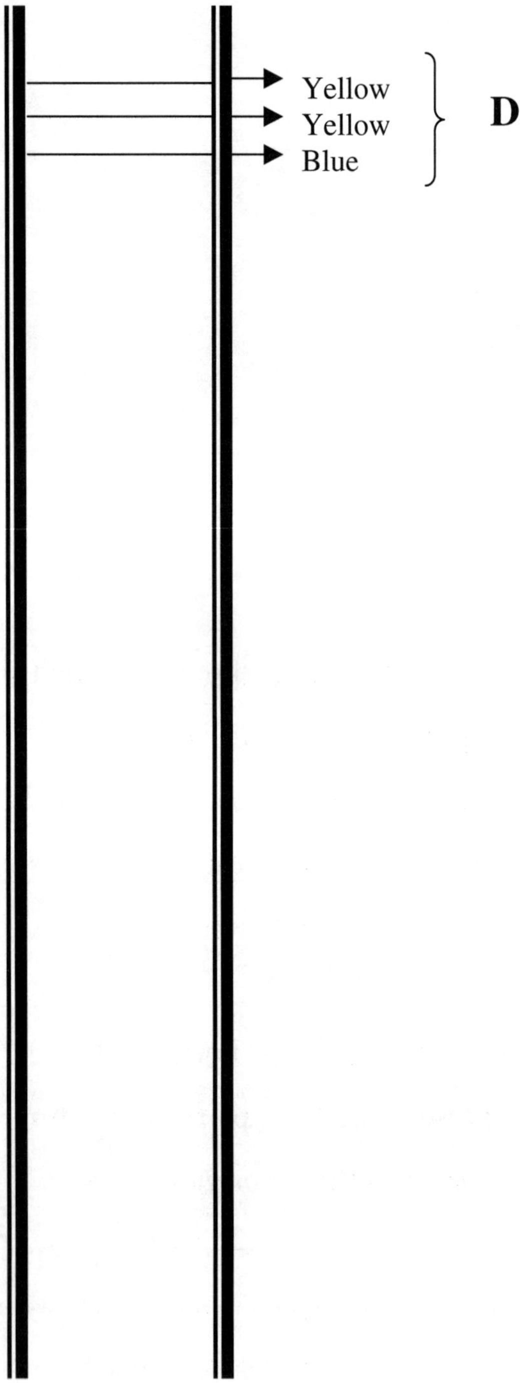

Third Step: Find out what the colors represent. Every three consecutive colors correspond to a letter in the alphabet (see the color letter codes above). For example, the first three colors listed in First Step are: yellow, yellow, blue - which corresponds to the letter D.

Follow the same three steps for the rest of the puzzles.

Puzzle Number 2

Red→green→yellow→red→blue→blue→red→yellow→yellow→yellow→yellow→yellow→red→yellow→yellow→yellow→red→red→yellow→green→blue→red→yellow→green→yellow→yellow→yellow→red→blue→yellow→yellow→red→red→red→blue→blue→red→green→yellow→red→red→blue→red→yellow→yellow.

Puzzle Number 3

Red→red→red→red→blue→blue→red→red→red→yellow→red→red→yellow→green→blue→red→yellow→yellow→yellow→blue→blue→yellow→red→blue→yellow→green→blue→red→blue→yellow→yellow→blue→blue→yellow→yellow→yellow→yellow→yellow →yellow→red→blue→yellow.

Puzzle Number 4

Red→yellow→yellow→yellow→red→green→red→red→yellow→red→red→blue→yellow→red→green→red→yellow→yellow→red→yellow→yellow→red→red→red→yellow→yellow→yellow→yellow→red→blue→yellow→red→blue→yellow→green→green→yellow→green→blue→yellow→yellow→blue.

Answer:

The encrypted sentences are as follows:

Puzzle 1. Dad has a big nose

Puzzle 2. Roses have thorns

Puzzle 3. Mom has flat feet

Puzzle 4. Skunks smell bad

Bonus:

- Create your own message.
- Create a 3-D model of the DNA molecule carrying your message. You can use different craft materials such as clay, colored toothpicks or Popsicle sticks. Use your imagination!

  A DNA model is described in the next section.

Making a DNA Model

Materials

Six different colors of Crayola® Magic Model in red, blue, green, yellow, white and black. (Magic Model hardens within 24 hours). Optional: both strands can be made with either the same or different colors.

Procedure

Take black and white Crayola® Magic Model and make two long, thick spaghetti-like strands. Lay them parallel. These will be the sides of the DNA ladder (DNA backbone). For each rung, you will use two colors corresponding to the base pairs. For example, adenine (red) will pair with thymine (blue) and cytosine (yellow) will pair with guanine (green). (See Figure 5)

Once you have completed the rungs, twist the ladder into the double helix. Twist the left strand under the right to form a right-handed helix. There should be ten base pairs between each full twist (See Figure 6).

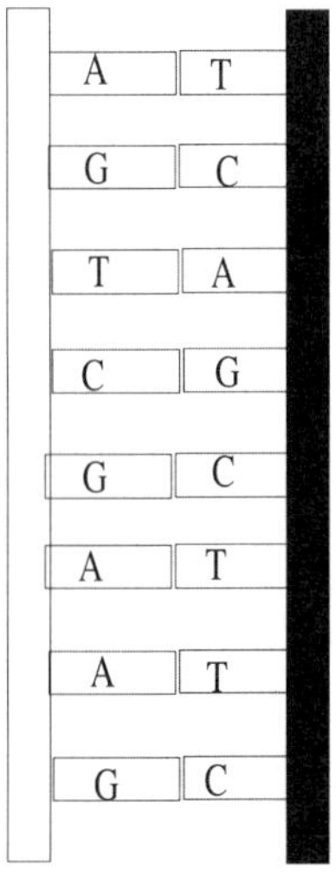

Figure 5

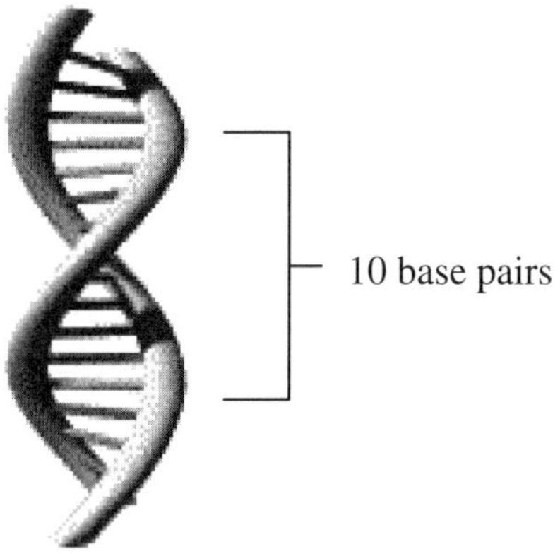

Figure 6

How to Extract DNA

DNA is the blueprint of life. It is found in all living things. The DNA molecule is six feet long and it is made up of three billion base pairs (in a haploid cell). It is supercoiled and tightly packed to fit the nucleus of a cell. You can extract DNA from either a fruit or a plant. The following procedure shows how DNA can be extracted from onions. Carry out your experiment with great care, as DNA (and especially RNA) can be easily degraded. Wear surgical gloves, if possible.

Materials

Medium-sized onion

Bowl filled with ice water

Strainer or coffee filter

Distilled water

A 100-ml measuring cup

Two glass containers

Toothpick or straw

Rubbing alcohol

Clear dish detergent or shampoo

Procedure

1. Chop an onion into very small pieces and put in a glass container.
2. Measure 90 ml. of distilled water into a second glass container (or cylinder). Add one teaspoon of salt and dissolve it in the water. Add 10 ml. of the detergent and mix.

3. Add the solution from the second container to the onion. Swirl several times and leave the mixture in a warm water bath (about 60^0C) for about 15 minutes. You can use warm tap water and if the temperature goes down, keep adding warm water to bring the temperature up again.
4. Next, place the DNA mixture in a bowl of ice for 5 minutes, and gently press the mixture on the side of the bowl with a spatula or spoon.
5. Filter the solution through a coffee filter over a glass container. The clumps will stay behind in the filter.
6. Add slowly equal volume of cold rubbing alcohol to the filtered solution.
7. The DNA (white color) will precipitate out from the solution. Sometimes it takes about 30 minutes.
8. Spool out the DNA by using a toothpick or a straw. Place the DNA in a tube or a small container and add rubbing alcohol. Store in the refrigerator or freezer.

Questions:

- What is the purpose of the detergent and the salt?

Answer: The cell wall and nuclear membrane are composed of fat. The detergent mixes with the fat and disrupts the membrane to expose the DNA. The salt is positively charged and will neutralize the DNA, which is negatively charged (because of the phosphate group). The DNA mixture will precipitate.

- Why filter the solution?

Answer: To get rid of all cell debris except for protein and DNA.

- What is the purpose of the alcohol?

Answer: DNA is insoluble in alcohol. When alcohol is added to a solution containing DNA, the DNA will come out of the solution (or precipitate out). Other cell components will be left behind.

How to Make a Cell Model

This is a very simple method for making a model of a cell. There are several web sites listed at the end of this book that have other cell models.

Materials

Play-Doh® or Crayola® Magic Model.

Procedure

Find a colored picture of a cell (Chapter 2). Make an irregular/round flat cell model with a white-colored Play-Doh®. This is the cytoplasm. Pick another color for the cell membrane. With different colored Play-Doh®, make shapes for nucleus, mitochondria, ribosome, endoplasmic reticulum, golgi complex and lysosomes. If you are doing a plant cell, make the chloroplast and add a cell wall. Try to make the correct shape of each organelle as much as possible. Stick all the organelles in the correct position in the cytoplasm.

Alternatively, you can use different craft materials for the organelles.

SECTION 3

EXERCISES

DNA Replication

Fill in the correct nucleotide for the complementary strand.

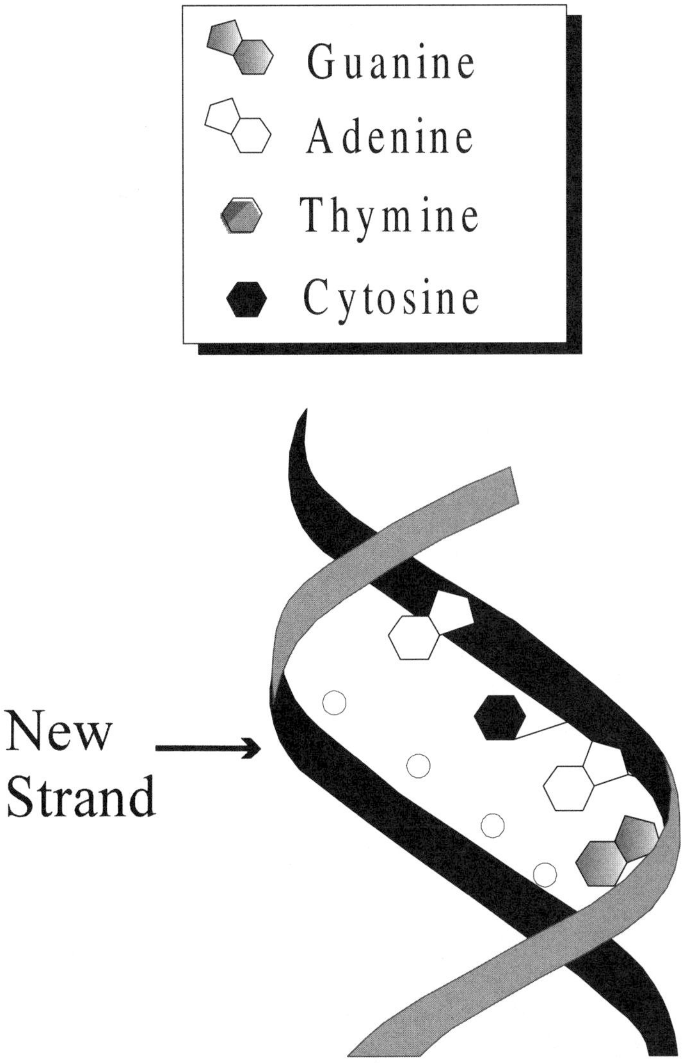

Answer:

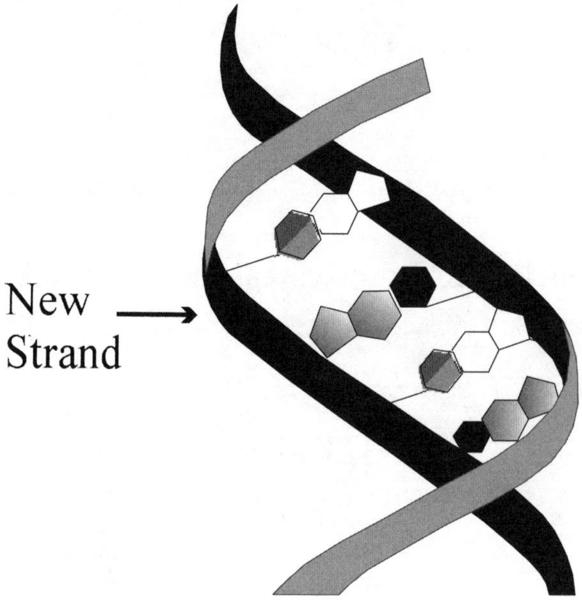

RNA Transcription

The transcription enzyme (RNA polymerase) binds the DNA and transcribes it into RNA. Fill in the correct nucleotides.

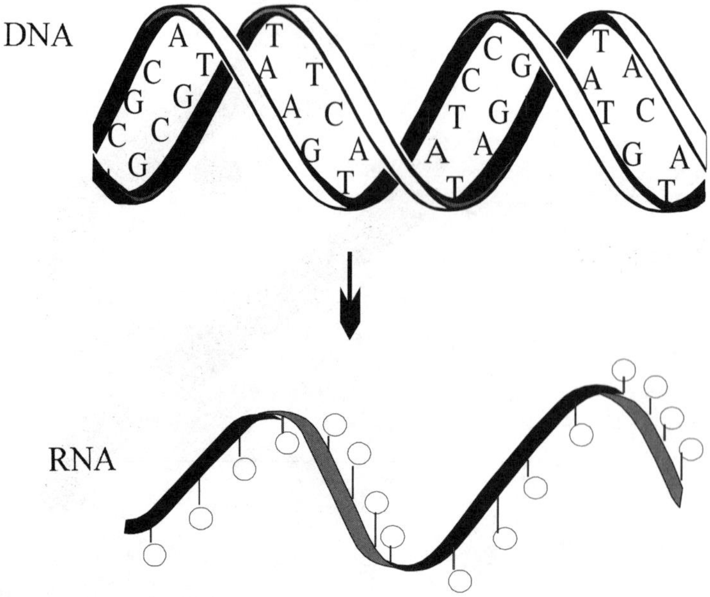

Answer:

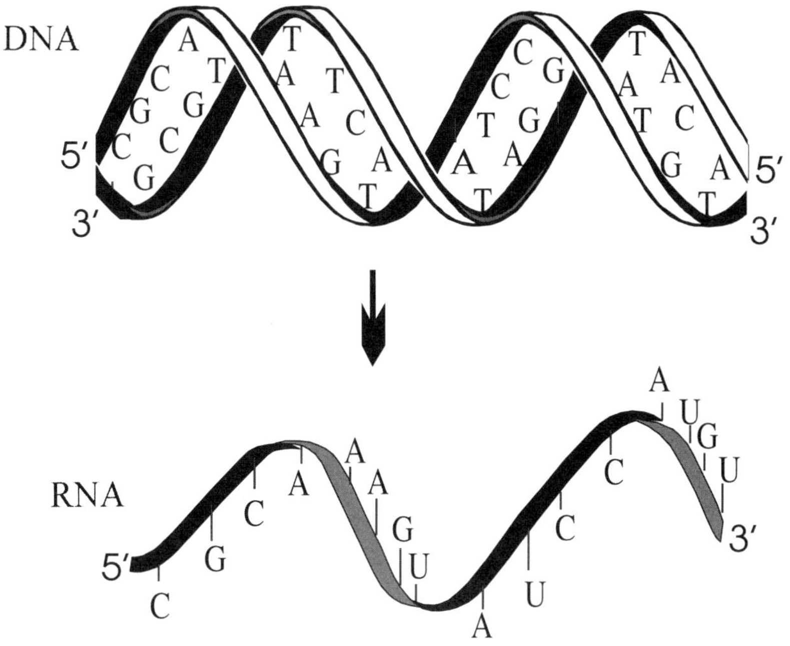

Mutation

A hypothetical scenario:

Dr. Smith discovers the gene for hair color. He locates its position to chromosome 2. A partial gene sequence for brown hair is analyzed. One of the two panels below represents the code for the brown hair trait.

A blue-haired graduate student joins Dr. Smith's team. She insists that she does not dye her hair but that she is born with the unusual hair color. She wants to learn to sequence her gene for hair color. After many failed attempts, she finally is able to sequence the gene from a sample of her blood. One of the panels below is from her DNA on chromosome 2.

1	2
A —— T	A —— T
T —— A	T —— A
C —— G	C —— G
C —— G	C —— G
G —— C	G —— C
T —— A	T —— A
T —— A	T —— A
C —— G	C —— G
A —— T	A —— T
T —— A	T —— A
T —— A	T —— A
A —— T	A —— T
T —— A	T —— G
C —— G	C —— G
G —— C	G —— C
G —— C	G —— C

Questions:

- Which sequence is the **wild type** and which is the **mutant**?
- Where is the mutation? Circle the incorrect pairing.
- If the mutation causes a new amino acid to be inserted, which is different from that of the wild type, what type of mutation would it be? (Types of mutations were not discussed in the book).

Answer:

This type of mutation is called misense mutation.

Wild Type | Mutant

1 | 2

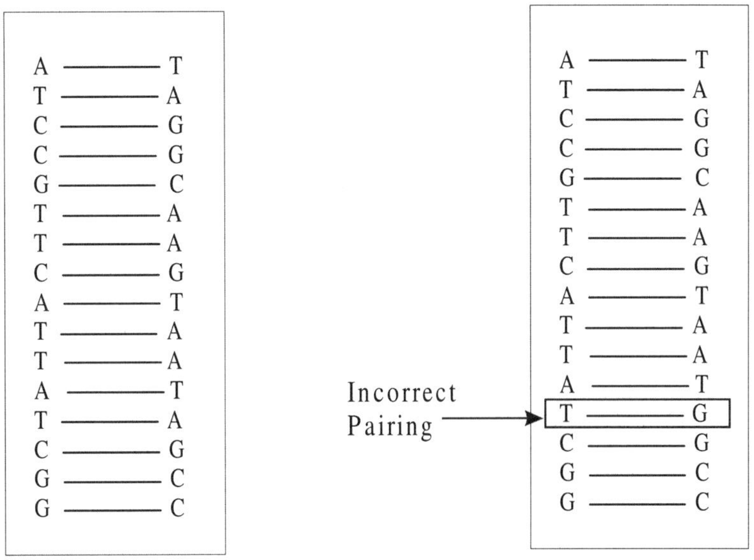

The Cell

Label the arrows for each part of the cell.

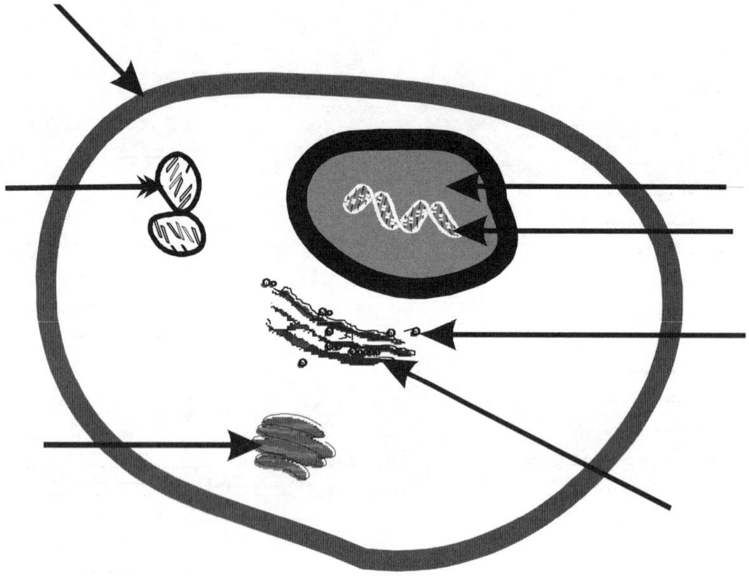

Answer:

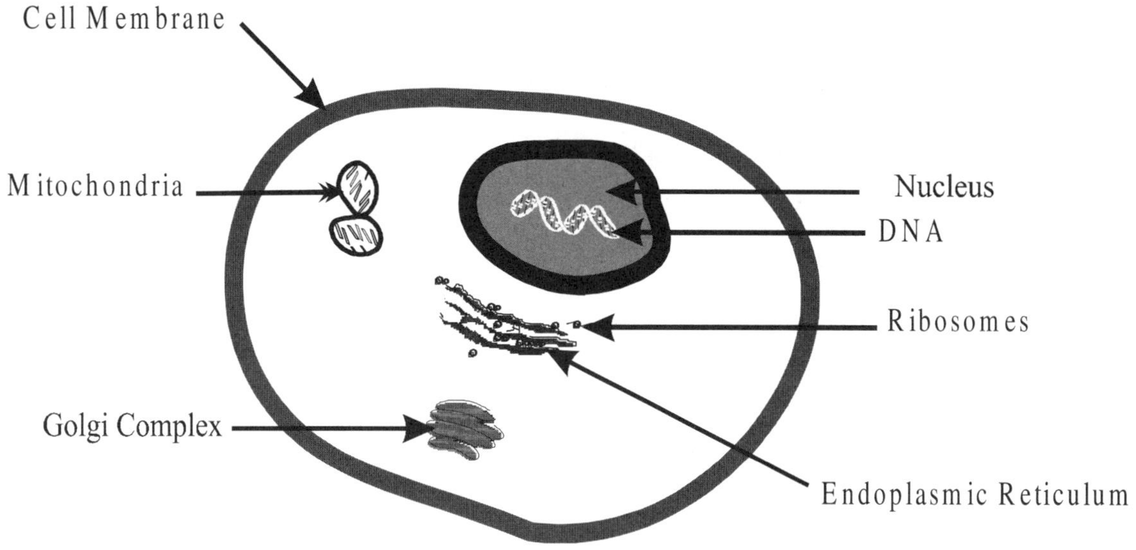

The Cell

Recommended web sites:

Genetic Science Learning Center
http://gslc.genetics.utah.edu/

DNALC Homepage
http://vector.cshl.org

The Human Genome Project
http://www.ornl.gov/TechResources/Human_Genome/home.html

The Gene School -Interactive
http://library.thinkquest.org/19037/interactive.html

The Biology Project
http://www.biology.arizona.edu/

Blazing a Genetic Trail
http://www.hhmi.org/GeneticTrail/

Science Magazine
http://www.sciencemag.org/

The Site for Health and Bioscience Teachers and Learners.
http://www.accessexcellence.org/

Other sites:
http://sciencematters.com/cloned/
http://www.cellsalive.com/
http://micro.magnet.fsu.edu/cells/index.html